A message to families...

This book is designed to be a conversation tool to help your family engage in discussion about when is the right time to receive a smartphone. This book includes two contrasting stories:

"Suppose You Give a Kid a Smartphone"
"Suppose You WAIT to Give a Kid a Smartphone"

This first story explores what it could look like if you give a kid a smartphone too soon. It isn't meant to frighten you, but sadly, this story describes a reality that is taking place all across the world and changing this generation.

Published by Engage Family Ministries
www.engagefamilyministries.org

ISBN: 979-8-9935870-0-4
Printed in the United States of America

First Edition

SUPPOSE YOU GIVE A KID A SMARTPHONE

By: Steve and Sara Otey

Suppose you give a kid a smartphone...

He'll probably want some
apps to go with it.

Once he has some apps,
he'll spend all day
playing on his phone.

3

He won't go outside or hang out with friends as much.

As the addiction grows,
he'll stay up late, missing out on sleep.

5

He'll be exhausted, and he'll have trouble focusing in class.

Once he gets social
media, he'll spend
hours and hours
just scrolling.

Constant notifications will fragment his attention,
and it will be tough for him to focus on anything at all.

As he scrolls, he'll compare his actual life
to everyone else's highlight reel.

He might start to feel anxious or depressed
and search for validation online.

The likes, comments, and follows might become more important than in-person interactions.

He might become obsessed with his phone
because of the dopamine it provides.

And when the online attention doesn't give him
what he hoped for, he might feel empty and alone.

And chances are, giving a kid a smartphone

might light up his face...
but may take away the light in his eyes.

If you've just finished reading the story on the other side of this book, you may be ready for a different approach. This story, Suppose You WAIT to Give a Kid a Smartphone, offers hope.

In his book The Anxious Generation, Jonathan Haidt advocates for new cultural "norms" to help children grow up healthier and happier. In this next story, we imagine what could happen when those four norms are lived out:

1. No smartphones before age 14
2. No social media before age 16
3. No phones in schools
4. More outdoor playtime for kids

Waiting to give your child a smartphone is hard—really hard. You'll feel like you're swimming upstream while other parents hand out phones in fourth grade. But waiting allows your child time to enjoy childhood, build strong face-to-face friendships, and mature in wisdom.

We wrote this story to give you hope—to show you what's possible when you choose courage over convenience.

Choose wisely. Choose time. Choose hope.

To the families
who want something different than
what this world has to offer.
May you be bold and courageous.

SUPPOSE YOU WAIT TO GIVE A KID A SMARTPHONE
By: Steve and Sara Otey

Suppose you wait to give a kid a smartphone…

She'll have more maturity to handle the responsibility that goes with it.

Once she gets her phone,
she'll use it with purpose and balance.

Her phone will help her connect
with friends in meaningful ways.

At night, she'll charge it outside of her bedroom.

She'll feel rested and stay focused in class.

When she uses social media,
doom scrolling won't be the norm.

Smart boundaries will protect her,
and she'll be able to focus on what matters most.

As she scrolls, she'll remember that
her worth comes from God, not from curated posts.

She'll find her joy in Christ,
and won't chase affirmation online.

Face-to-face moments will matter
more than likes, comments or streaks.

Because she has a heart of wisdom,
she will know how to manage her phone.

Knowing her phone can't truly satisfy,
she'll turn to God, family, and friends.

Chances are, she'll be thankful
you waited to give her a smartphone,

because now she's ready to change the world...
not just scroll through it.

We hope this book inspires you to make informed and wise choices about smartphone use. The outcome of this second story, however, cannot be achieved simply by waiting longer to give your child a phone. It takes intentional parenting that teaches, models, and trains your child's heart to develop a healthy relationship with technology.

Parents, may you faithfully model what it looks like to use technology for God's glory, engaging your children in meaningful conversations along the way. We pray the Lord will direct your steps and make your path straight as you trust Him with every parenting decision.

Steve and Sara Otey
Engage Family Ministries

EngageFamilyMinistries.org

The story doesn't have to end like this. Flip the book over and read "Suppose You WAIT to Give a Kid a Smartphone."